PET CATFISH

The Ultimate Guide to Catfish Farming and Keeping

Raising Catfish as Pets, Aquarium Care, Nutrition, Health, Breeding, and Tank Setup

Edna R. Francis

PET CATFISH

All rights reserved. No part of this publication may be reproduced, distributed, or transmitted in any form or by any means, including photocopying, recording, or other electronic or mechanical methods, without the prior written permission of the publisher, except in the case of brief quotations embodied in critical reviews and certain other noncommercial uses permitted by copyright law.

Copyright © Edna R. Francis,2024.

TABLE OF CONTENTS

CHAPTER ONE 5
  INTRODUCTION 5
    1.1 INFORMATION ABOUT CATFISH 6
  1.2 FACTS AND OVERVIEW 7

CHAPTER TWO 10
  2.1 HABITAT OF THE CATFISH 10
    2.2 CATFISH POPULATION DISTRIBUTION 12

CHAPTER THREE 14
  3.1 BEHAVIOURAL AND PHYSICAL CHARACTERISTICS 14
  3.2 BEHAVIOUR OF CAT FISH 15
  3.3 CATFISH VARIETIES AND THEIR BASIC CARE 16

CHAPTER FOUR 50
  4.1 DIET AND NUTRITION 50
    4.2 HUMAN-CATFISH INTERACTION 53
    4.3 DOMESTICATION 54

CHAPTER FIVE 56
  5.1 BASIC CAT FISH CARE AND REQUIREMENTS 56
  5.2 BREEDING AND REPRODUCTION 58

CHAPTER SIX 60

   6.1 HEALTH PROBLEMS AND TREATMENTS 60

      6.2 SIGNS OF A HEALTHY CAT FISH 62

CHAPTER SEVEN 63

   7.1 INSTALLATION OF A CATFISH TANK 63

CHAPTER EIGHT 68

   FAQ AND ADVANCED TIPS 68

CHAPTER ONE

INTRODUCTION

The Catfish is a well-known animal that may be easily identified by its fleshy barbels that resemble whiskers. These fish are classified as belonging to the taxonomic order Siluri formes. Within the group, researchers have identified over 30 different families, which contain thousands of distinct species. Catfish may be kept as pets, and they make interesting fish for aquariums at home. There are several kinds, sizes, and color options, so you can frequently find the ideal catfish to match your current arrangement. These aquatic creatures may either be kept in their own tank or added to a communal tank where they can be quite helpful in cleaning the food and debris that settles to the bottom of the aquarium. To make sure you have all the information you want if you have questions regarding the specifics of keeping catfish as pets, we have put together this helpful guide.

1.1 INFORMATION ABOUT CATFISH

Despite the fact that these fish vary greatly in size, color, and form, you will probably be able to identify one when you encounter it. Researchers refer to the numerous long, fleshy growths that are located close to their lips as "barbels" because they resemble cat whiskers. These fish come in an exceptionally broad variety of sizes. While some are only a few inches long, others can reach eight feet or longer.

1.2 FACTS AND OVERVIEW

These fish exhibit a wide range of characteristics and behaviors due to the astounding variety of species. Find out here what makes a few particular species special.

The Mekong Giant Catfish This species, which is very uncommon and enormous, is in danger of going extinct. This species is classified as Critically Endangered by the IUCN as a result of overfishing and habitat degradation. The longest individual ever reported was 8 feet, 10 inches long, making this species the biggest exclusively freshwater fish species.

CANDIRU - If you are eating something, finish it before you read any more since you could become satiated thereafter. This little species, commonly referred to as the vampire fish, has a bad reputation. According to reports, individuals have seen this parasite fish swim multiple times.

According to reports, this parasite fish has been seen swimming up swimmers' urethras multiple times.

Imagine putting your hand in a hole beneath a log while swimming in a murky river, only to have something snatch it up and put it in its mouth! Although it may sound like a nightmare, this method of hand-catching this species of catfish is called "noodling"!

The Giraffe Catfish may be the most alluring-looking animal you've ever kept in a fish tank! Make sure you have the resources and know-how to take care of this species, which may reach lengths of two to three feet and is rather unusual.

CHAPTER TWO

2.1 HABITAT OF THE CATFISH

These fish may be found in a staggering variety of settings. They inhabit a huge variety of habitats, some of which appear almost uninhabitable.

They live in a variety of environments, some of which are rivers, lakes, ponds, streams, swamps, and more. While the majority live in a variety of environments, some do prefer certain ecosystems or geographical areas.

The majority of catfish are bottom-dwelling species. While many catfish may spend some time swimming in the middle of the water column, others utilize their barbels to look for food that has been buried in the bottom. Actually, their barbels are packed with taste buds that are always on the search for delectable treats. Your catfish will demand bottom access in light of this. Except for thickly planted tanks, the majority of aquarium configurations are suitable for catfish. Catfish are easily lost and entangled in dense undergrowth.

Some species, like plecos, prefer softer wood to chew on. Make sure the wood is suitable for aquariums before putting it to a tank. Some types of wood might release harmful compounds into the water of your aquarium. Contrary to popular belief, catfish are omnivores, with certain species favoring herbivory while others favor carnivory. Keep in mind that the family of catfishes has roughly 3000 different species. Expecting most catfish to remove your algae is unrealistic. In the majority of freshwater systems, you will need to regulate your algae by restricting its nutrition source.

2.2 CATFISH POPULATION DISTRIBUTION

Wait till you see their range if you believe their size varies! All continents, with the exception of Antarctica, have freshwater environments and coastal areas where you may locate this fish.

They are found on numerous nearby islands as well as in Eurasia, Africa, the Americas, and Australia. Some people only exist in a solitary area or perhaps a single river system, while others dwell throughout extraordinarily broad geographic regions.

PET CATFISH

Catfish are the most powerful river predator. In the wild, the greatest weight can reach 500 pounds.

Catfish hunt at night. They like to swim towards the thickets at night to look for prey.

Catfish are very intelligent and cunning creatures. They wiggle their mustache to imitate worms and keep their mouth open to attract food.

Female ancestors can change sex if there is no male in the pond.

Catfish like to stay at home. They spend their entire lives in a pit that they "build" for themselves.

CHAPTER THREE

3.1 BEHAVIOURAL AND PHYSICAL CHARACTERISTICS

Catfish are present in every continent (with the exception of Antarctica). They can survive in a broad range of environments and can be found in almost all freshwater rivers.

This undersea dweller is regarded as the most robust and modest fish species, making them ideal candidates for home aquariums: they eat nearly everything, are quiet, and are not demanding of their environment. The one big downside is that many catfish may grow to be quite enormous.

The absence of scales, which is typically replaced by bone plates or thick skin that secretes a lot of mucus, distinguishes these residents. The fish are frequently awkward and sedentary as a result of this cover. Another distinguishing trait is the existence of antennas, which range from two to four pairs. which can be seen on both sides of the mouth the antennae can be branched or changed into lips depending on the species.

3.2 BEHAVIOUR OF CAT FISH

As is the case across in most other sections, behavior, both social and individual, changes based on the species at hand. Some have nocturnal behavior and spend their nights hunting for prey. Other species feed during both the daytime and the nighttime, or during the day only.

Socially, some species live in groups, while others live alone. Some species also claim and defend their own territories or burrows.

3.3 CATFISH VARIETIES AND THEIR BASIC CARE

Certain catfish species are better suited to aquarium living for a variety of reasons, including water conditions and ease of maintenance. We've produced a list of the top 10 possibilities for your next tank purchase:

1. Pleco bristlenose

This bottom-dwelling Catfish species is often known as a "living vacuum cleaner" for your tank because to its algae-eating activity and placid nature.

This Catfish species will spend its time vacuuming the substrate of your aquarium, clearing it of leftover food and algae. Bristlenose Plecos get along nicely with other calm freshwater species and thrive in community tanks of at least 45 gallons or more.

Bristlenose Plecos are a species for novices because of their easygoing attitude, unusual appearance, and resilient nature. This Catfish species is still tiny in comparison to other Catfish, averaging 4-5 inches in length. They have a distinct look that makes them easily identifiable.

The Bristlenose Pleco is a unique Amazon catfish. This fish is regarded rare and may become true aquarium jewelry. It is critical not to mix it up with other varieties. Its native home is the Amazon River in South America. This fish's body is black and covered in tiny stars. It poses no threat to neighbors of comparable size and disposition. It gets along nicely with its non-aggressive neighbors of similar size. Two barbules, ternetii, and scalarii.

The common pleco is the "suckerfish" that is frequently observed attached to aquarium glass. It is marketed in pet stores as three or four-inch juveniles but can grow to be a couple of feet long as an adult. This fish is obviously not suitable for any but the largest home tanks. Consider 150 gallons or more.

The common pleco is an avid algae eater, and if you have a large enough tank, it may help keep things clean. To ensure that it has enough to eat, supplement its diet with algal wafers. It's also a good idea to have some driftwood for your pleco to rasp on, as well as some dark caves for hiding.

Large plecos can be hard on living plants and can grow aggressive as they age. Otherwise, if you have a large enough tank, there are various lovely pleco kinds to pick from. Simply ensure that you properly examine any kind you are interested in to ensure that you can accommodate them as adults.

Their broad head is "decorated" with thick tentacles, giving them a distinctive look and earning them the moniker Bristlenose.

Their body is flat and brown, green, or gray in hue, with yellow or white markings. It's not a terribly attractive fish, but it's really helpful and mellow.

Here are a few things to consider if you want to keep Bristlenose Plecos:

- **They eat plants:** Bristlenose Plecos prefer an herbivore diet, so offer them spirulina wafers, algae, flakes and granules, as well as soft foods.
- **Avoid tankmates who are aggressive**: Many calm fish are excellent companions for this Catfish species; however, aggressive tank mates should be avoided.
- **Make hiding places:** Because they are generally nocturnal, give adequate hiding places (driftwood, caverns, boulders) for them to retreat during the day.
- **Bristlenose Plecos are easy to care for:** thrive in a range of tank settings, and are friendly fish that

may be kept in a communal tank with other peaceful tank mates.

2. Gold Nugget Pleco

There are several varieties of the stunning aquarium fish known as the Gold Nugget Pleco. **Habitat**: South America, Venezuela, and the Amazon River.

Except for other Pleco, they get along well with the majority of other fish species. They are omnivorous but consume largely algae that grows on driftwood and rocks. Additionally, specialty food and seaweed flakes can be offered to these fish.

These Plecos are my second favorite species of Catfish. Their name alludes to their beauty: a black body speckled with pale yellow dots. Yellow is also used to tint parts of their fins.

Their size ranges from 6 to 9 inches, hence they demand larger tanks due to their size. A minimum tank capacity of 55 gallons is recommended.

They're also a little more expensive, but they're absolutely lovely animals that make it all worthwhile.

The Gold Nugget Pleco eats a range of meals, including algae wafers, fresh veggies (zucchini and cucumber), and algae that grows naturally on driftwood in your tank. They love grazing on plants as well, but not to the extent that certain herbivorous animals do.

When selecting Gold Nugget Plecos, keep the following aspects in mind:

- **Select fish that like the uppermost parts of the tank**: Gold Nugget Plecos won't be bothered by fish that remain in the middle to upper levels of the aquarium. Males of their own species or even other males have been observed to be caught by Gold Nugget Plecos;

- **Give out dishes with a lot of vegetables in them:** Although they are classified as omnivores, Gold Nugget Plecos also eat a lot of plants;
- **Keep an eye on the water's conditions:** High 70s water temperature is preferred by them.

> **Even though they are a little more expensive and aggressive, Gold Nugget Plecos may make a fantastic addition to a community tank:** They can make a fantastic view to a community tank long as they are housed with fish that get along with them.

3. Corydoras Catfish

With goldfish and other viviparous species, the Corydoras is one of the most well-liked tropical aquarium fish. They are native to South America's southeast, namely the area of the La Plata Estuary, where freshwater reservoirs are found. Greek terms Corys and Doras are the roots of the word Corydoras. The first is frequently interpreted as "helmet," and the second as "skin." However, a protective cover or piece of armor is a more realistic definition.

The little cory catfish is one of the most well-liked bottom-dwelling fish for aquariums with tanks greater than 10 gallons. They are little, quiet, cautious fish that only reach lengths of a few inches. They spend much of their time rummaging over the substrate in search of food. They frequently conceal themselves in plants or decorations when they are motionless.

Cories are schooling fish, thus to reduce stress, they should be maintained in groups of six or more. They require little maintenance and should be given sinking pellets in addition to the food they would naturally scavenge.

CORY CATFISH VARIETIES

There are several cories available for the home aquarium. Here are a few common cories seen at pet stores:

- Bronze Cory
- Albino Cory
- Pepper Cory
- Pygmy Cory
- Bumblebee Cory
- Panda Cory
- Sterba's Cory
- False Julii Cory
- Skunk Cory

The emerald green cory, also known as the emerald brochis, is another "cory" you may come upon. These critters get a little bigger than real cories, but they're still little fish, max out around three or four inches. If you have a tank large enough to accommodate many different schools, avoid mixing and matching kinds. Keeping them alone or in small groups might create health problems and alter their disposition, making them even more cautious and less active.

Because they are bottom dwellers who spend a lot of time scavenging in the substrate, there is a danger of harm to their delicate bellies if housed on gravel substrates; as a result, there are several particular needs you should be aware of:

- Provide a sand substrate: gritty gravel and other sharp surfaces might hurt their barbels and underbellies, therefore sand is preferable.
- Stable water parameters: Although Cory Catfish are a resilient species that is more forgiving of water conditions, they

nevertheless like a stable tank environment.
- Corys are an omnivorous species, which means they require a balanced diet of meat-based and plant-based meals.
- Corys should not be kept with aggressive tank mates since they are too placid to defend themselves when provoked. Also, be careful not to overfeed them!
- Corys may live between 12 and 15 years if properly managed for, so make sure you're up to current on their needs.

4. Inverted Catfish

In the wild, this fish may be found in Africa, in the basin of the Congo River. It consistently swims up with its belly, which is its major characteristic. There is still no explanation for this behavior. This fish has a tough, slimy coating covering its somewhat flattened body on the sides. You should select aquarium fish with similar size and temperaments for your neighbors.

When it comes to unusual fish, the Upside-Down Catfish is one that is sure to turn some attention, but not necessarily for its beauty.

Upside Down Catfish, as the name suggests, spend a lot of their time swimming...well, upside down. This isn't to say they only swim like this; you'll also see them swimming properly, especially while foraging in the substrate. Despite their unusual swimming style, they have a striking similarity to Cory Catfish in disposition and even look.

They prefer to be housed in groups of at least 6, perform best in bigger tanks of 30-45 gallons, and are an omnivore fish.

Nevertheless, they get along well with other calm fish like African Tetras and make good community fish. You can sure that because they are omnivorous, they will consume little fish that fit in their jaws.

They favor planted aquariums, ideally with broad-leafed plants since they like to browse the undersides of leaves.

Additional Details on Upside-Down Catfish Include:

- ➤ They require suitable hiding places like caves, driftwood, and rocks to conceal during the day because they are more active at night;

- Easily adjusts to any kind of food: This species will consume any form of food, including live, dried, and frozen items, and will gladly graze on algae, which they can do so without difficulty;
- Keep it away from large predatory fish: Large predatory fish will try to eat Upside-Down Catfish; however, the fish can become trapped in the eater's throat because the Upside-Down Catfish will erect its spines when threatened.
- This Catfish species might be a great addition to your community aquarium if you're searching for something truly "upside down" for your tank.

5. Glass Catfish

A peculiar aquarium fish is the glass catfish. This species is distinguished by its total lack of pigmentation, which renders the body of the fish translucent, allowing you to view its internal organs and even the aquarium backdrop. Because of their sensitivity to nitrate levels, glass catfish Consequently, you can only introduce fish into an aquarium that has a stable biological balance. It's also important to keep in mind that this fish is shy and should only be kept in groups with calm neighbors.

The Glass Catfish is a wonderfully distinctive species that will stand out in any aquarium. It's an impossible fish to miss because of its translucent body, which makes the fish's viscera and spine apparent.

Their absolutely translucent body is due to a lack of pigmentation, so much so that even their muscles appear light. They're also known as Ghost Catfish because to their unusual body. These fish make excellent communal fish, but because they are so sensitive, they must be kept with tranquil companions. Because they are sensitive to water factors, this species favors well-established aquariums. Because they are schooling fish, keeping them in groups of 5 or 6 is the best method to protect them from becoming reclusive and shy.

This species, like other Catfish species, has barbels on its top lip that are extended.

Glass Catfish are more demanding than the other Catfish species I've described so far, and I don't suggest them for novices.

Things To Keep An Eye Out For When Maintaining Glass Catfish:

- They're low light fish, which means they don't like bright lights and should have plenty of hiding spots and covering in a communal tank.
- Water conditions that be stable: Because they are sensitive to changing water conditions, maintain things consistent (they prefer slightly acidic water and temps in the upper 70s).
- Omnivorous species: Provide a well-balanced food rich in high-quality flakes and freeze-dried blood worms.

Glass Catfish are an intriguing species that I suggest only to experienced aquarists who can provide the ideal tank conditions for this fish.

6.. Pictus Catfish

Many freshwater aquariums can benefit from the Pictus. Due of its nocturnal habits, this creature needs dark illumination or cover throughout the day. They have an upper limit of 5 to 6 inches. The Pictus may coexist peacefully with other night fish in South America's Amazon and Orinoco rivers.

The Pictus Catfish, with its distinctive silvery body speckled with black spots and large barbels, is a hobbyist favorite. The Pictus Catfish is a popular among hobbyists because to its distinctive silvery body covered in black spots and large barbels. This kind of catfish also schools, so it's ideal to keep them in groups of five or six.

Because these fish tend to exhibit aggressive tendencies as they become older, be on the lookout for any symptoms of hostility.

They may be housed in a tank with other fish that aren't too tiny for their jaws. Finding tank mates that are the right size shouldn't be difficult because they often stay under 5 inches.

A combination of high-quality sinking pellets, vegetables, frozen bloodworms, and fresh brine shrimp works well. Because they are an omnivorous species, they will benefit from a diet that includes both meat and vegetable stuff.

This is a schooling fish, like many other Catfish species, therefore keep them in groups of 5 or 6.

However, keep an eye out for signs of hostility, since these fish tend to become aggressive as they age. They may be housed in a communal tank with fish that aren't small enough to fit their jaws.

They typically stay under 5 inches in length, so finding size-appropriate tank mates should be easy.

Because they are an omnivorous species, they will benefit from a diet that includes both meat and vegetable matter - a combination of high-quality sinking pellets, vegetables, frozen bloodworms, and fresh brine shrimp works well.

7. Catfish, Otocinclus

In Southeast Brazil, Otocinclus inhabits slow-moving rivers, lakes, and ponds that are choked with vegetation. It is the family's tiniest member. The fish has an extended body that is flattened in the belly and chest regions from top to bottom. Fish called Otocinclus are calm and modest in nature. It must remain with a group of at least five. Small and non-aggressive neighbors are ideal. It is wise to bear in mind that otocincluses clean the aquarium, thus it is not a good idea to keep them with fish that consume algae.

In terms of tank mates, the Otocinclus Catfish gets along well with other community tank species, particularly Cory Catfish, as well as several shrimp and snail species such as Mystery Snails,

Nerite Snails, Amano Shrimp, and Ghost Shrimp.

They have a considerable following in the aqua scaping community because of their voracious hunger for algae, unique characteristics, and placid temperament.

I don't really advise them to a newbie aquarist because they demand densely planted aquariums, especially considering that building up a planted aquarium may be pricey.

They need a planted tank with plenty of shelter and hiding spots, like caves and driftwood, in addition to sufficient water movement.

The Otocinclus Catfish is generally easy to maintain for; they will consume algae off all surfaces in the tank, including the plants, without harming the plants. Algae wafers and soft vegetables like zucchini slices should be included to their diet to round out their nutrition.

When keeping Otocinclus Catfish, keep the following in mind:

- Keeping them company with tank friends their size: Because of their modest size (around 2 inches), keep them in tanks with other fish of comparable size.
- They eat plants: Make sure to include foods high in vegetable matter in their diet.
- Avoid situations that are stressful: Maintain a stress-free habitat by maintaining a consistent tank state.
- Oto Catfish prefer to be housed in groups of 5-10, and because they are energetic swimmers, they

require considerable swimming room, so pick a tank that is at least 25 to 30 gallons in size.

8. The Plecostomus clown

This submerged creature is native to the Orinoco River in Venezuela and Colombia. Their dark hues help them blend in with the vegetation and debris around riverbanks, where they are typically found. They require an aquarium of at least 20 to 25 gallons despite their small size since they eat on wood and produce a lot of waste, which may rapidly ruin a small tank.

Because of its small size, the Bumblebee Catfish is a perfect complement to 20-30-gallon aquariums (up to 4 inches). It was given its name because of its yellow-black patterning.

This Catfish species is nocturnal, but if it smells food nearby, it will emerge from its hiding place during the day. It loves having a tank with a variety of hiding places where it may rest during the day.

Despite its small size, it has a big mouth and can consume larger items such as sinking pellets, frozen and freeze-dried food, and frozen and freeze-dried food. It can also eat small enough fish to fit in its mouth.

Males are known to get territorial when other males are around and prefer to stay at the bottom or side of the tank.

In addition to providing them with a fair amount of driftwood, there are a few other considerations while maintaining clown plecos:

- Maintain them with fish that won't consume them: Clown Plecos coexist with a variety of species, although big fish will devour them;
- Ensure the tank is at least 20 gallons;
- They are challenging to breed: Expecting to be able to breed Clown Plecos will be difficult.

If you want something different, clown plecos, which have an amusing appearance, might be a nice addition to a planted aquarium.

9. Catfish Bumblebee

The rivers of Thailand, Laos, Cambodia, and Vietnam are its native habitats. In slow-moving waterways, it coexists with roots of buried trees. During the day, it stays near to shelters, and at night, it ventures outside in quest of food.

Adults grow to be about 5 inches long. Dark brown dominates, with a number of golden vertical and diagonal lines. A single catfish needs an aquarium that is at least 25 gallons in size.

The Bumblebee Catfish is an excellent addition to 20-30-gallon tanks due to its modest size (up to 4 inches). It was named from its yellow-black patterns.

This Catfish species is nocturnal, but it will emerge from its hiding location during the day if it detects food nearby. It appreciates having a tank with several hiding spots where it may retire during the day.

Despite its tiny size, it has a large mouth and can eat larger foods such as sinking pellets or frozen and freeze-dried food. It can also consume fish that are tiny enough to fit in its mouth.

This particular type of catfish like to hide in caves, behind vegetation, holes in driftwood, and crevices in rocks at the bottom of the aquarium.

Should you decide to keep Bumblebee Catfish:

- Avoid eating snails and shrimp: Anything that the Bumblebee Catfish can fit into its mouth is a possible target for them;

- Hefty eaters: Feed them a combination of meaty feeds like brine shrimp or bloodworms and sinking pellets.
- The Bumblebee Catfish is a popular species because of its small size, calm temperament, and distinctive patterns.

10. Whiptail Catfish

These submerged creatures are found in Mexico and Venezuela. The fish loves oxygen-rich, crystal-clear water. Sandy soil and rivers with strong currents are where you may find it. It hides during the daytime in vegetation or caves. It occasionally leaps out of the water at the same moment. It is active at dusk and at night and feeds on algae and tiny water creatures.

Which catfish varieties are ideal for tanks with others?

Certain catfish species are more compatible with cohabiting an aquarium with other fish than others. It's crucial to buy a catfish that gets along with others if you currently have a tank with several fish or if you want to set up a mixed aquarium. The top four species for communal tanks are shown below:

- ❖ Pleco Zebra

    This fish is cautious and easily frightened by too energetic fish. However, this species is regarded as one of the most attractive. Adults grow to be around 3 inches long. The body design is made up of contrasting black and white stripes that seem like zebra stripes.

- ❖ Corydoras Catfish

  Corydoras are social fish, therefore keep them in an aquarium with 4-6 other species. They get along well with other species in a shared aquarium. These tranquil fish primarily swim in the water's bottom layers. The presence of six antennae grouped in pairs distinguishes this species of catfish.

- ❖ Pleco bristlenose

  These fish offer no harm to their neighbors of comparable size and disposition. They get along with their non-aggressive neighbors of the appropriate size. This fish's aquarium should include greenery and resting areas for the fish.

- ❖ Plecostomus the clown
  Individual specimens found in nature can grow to be 20 inches long, although the average length in aquariums is little more than 11 inches. Young plecostomus-fish are quite calm. They may be kept in a shared aquarium with other species as well as their own type.

CHAPTER FOUR

4.1 DIET AND NUTRITION

You could see a pattern here, however because to the enormous variety of species, you cannot categorize all fish as belonging to one kind of diet. Some people consume other animals and have largely carnivorous behaviors. Others are omnivores because they consume both plants and animals.

Depending on the species, hunting tactics also vary. Many hunts by ambush, waiting for prey to get too near before attacking. Others engage in active hunting. They hunt for tasty fish, crabs, insects, frogs, and other animals with their delicate barbels.

When maintained in an aquarium, catfish enjoy sinking pellets or large flake feeds. Catfish are benthic animals, which means they reside at the bottom of the tank or on the ocean floor in the wild. They dwell on the substrate, sifting through it in search of food. This implies that any tank food that floats on the water's surface is unlikely to be consumed by the catfish, perhaps leaving them hungry. Instead, choose food that sinks, allowing them to eat whatever the fish higher up in the tank drop or leave behind. Catfish should be fed on a lighter-weight substrate. For the majority of catfish, standard aquarium gravel to medium sand is a decent option. Smaller catfish may have trouble with bigger boulders, and larger catfish have been known to ingest rocks.

The majority of catfish species are nocturnal and prefer to be fed at night when the tank lights are turned off. Catfish frequently consume brine shrimp, Mysis shrimp, and blood worms, both live and frozen. If you choose frozen food, make sure it is completely thawed before adding it to the tank. Other eating habits vary according to catfish species.

If there is not enough room or food for the different fishes in the tank, the catfish may consume them. Catfish are prohibited from living in the same tank as goldfish and some cichlids. Guppies, scalars, barbuses, tetras, neons, and rainbows can all be used as hooks for catfish.

4.2 HUMAN-CATFISH INTERACTION

People deal with these fish in a variety of ways all throughout the world. Numerous species, many of which are bred and raised in fish farms, are among the hundreds of species that people consume as food. Some species are also kept by people in aquariums in their homes.

These fish are affected to varying degrees by overfishing and other human activity. Some species, like the Mekong Giant species are gravely threatened by human activity. Because there are so many other species, human activities have little effect on them. Every species has unique differences.

4.3 DOMESTICATION

Pick your species of catfish wisely before selecting one for your aquarium. Here is a useful reference to the catfish species most suited for aquariums in homes. In order to keep your catfish happy for the duration of its life, it is essential that it be an adult size. It might be hard to imagine how big fish might get when they all start off the same size, and not all pet retailers will provide you this information. Even selecting a placid plecostomus can be difficult because some species only reach a maximum length of 8 inches while others can reach lengths of many feet. Additionally, certain catfish species, like the Corydoras species, are calm bottom feeders that get along well with other tiny fish. whereas other predators, like the Redtail Catfish, will consume any other fish that can fit in their mouths! To ensure that the catfish you are bringing home can fit into your aquarium given its size and the other species you already have, it is imperative that you conduct prior study. Don't buy anything if you can't be sure the species, you're looking at is the right one.

CHAPTER FIVE

5.1 BASIC CAT FISH CARE AND REQUIREMENTS

Different species have different demands. Some require larger tanks as they grow larger than others. Others are good additions to tiny tanks since they only grow to a length of a few inches. All love frozen or live food, but the majority thrive on suitable professionally prepared fish diets.

Like all other fish species, catfish require high-quality water and a balanced diet to maintain optimum health. Keep in mind that not all catfish enjoy munching on algae, so be sure you feed them frequently. Catfish kept in tanks with other species will do just well on the majority of tropical fish diets. You should choose a diet closer to the herbivore or carnivore end of the range if your aquarium is only for catfish. Catfish frequently forage throughout the day, so anticipate them to consume some of the food you put in your tank. You will still need to perform your routine maintenance even if catfish are performing the cleaning.

The majority of catfish live in groups. They must share the tank with five or six other fish. A capacity of 20 gallons is appropriate for such a business. The ideal aquarium temperature for catfish is around 70 °F. Catfish can live in water with an acidity of 6-7 and a hardness of 1-17. Aeration and filtration should be present in the aquarium. It is also required to replenish the water half once a week.

It is preferable to keep a catfish aquarium in the house's quietest room. Catfish are quiet creatures, and loud noises make them anxious. Loud music or slamming doors might have a detrimental impact on their well-being. You cannot place the aquarium in a bright area since these fish like obscurity. A through-and-through aquarium is not appropriate; the back wall must be turned to the wall or adorned.

5.2 BREEDING AND REPRODUTION

Reproductive behavior varies depending on the species under discussion. Every one differs in terms of mating behaviors, egg production, incubation times, and other factors.

The group as a whole typically reproduces through spawning. The female releases her eggs during spawning, and the male fertilizes them externally.

CHAPTER SIX

6.1 HEALTH PROBLEMS AND TREATMENTS

Aquarium catfish, like other fish, are susceptible to a variety of illnesses. Diseases can arise as a result of catfish's low content, which can come from other fish, live food, plants, and snails in the tank.

In the first scenario, you may try to remedy things by changing the water more frequently and improving the confinement circumstances; in the second case, you require medicine. Do not pour the medications into to the tank immediately after diagnosing the fish and giving therapy. The problem is that aquarium catfish do not tolerate treatment with salt or copper-containing preparations.

Beginners are frequently concerned with a momentary shift in the color of catfish (paling or light spots). This is not an illness, but rather a stress reaction. Catfish may turn pale or blend in with the surrounding environment. When the fish calms down after a while, the natural color will be restored. Otherwise, aquarium catfish are treated in the same way as other fish are. Catfish will be significantly less sick if they are kept in large aquariums with live plants, excellent aeration and filtration, and weekly water changes of 15-25% of the volume.

6.2 SIGNS OF A HEALTHY CAT FISH

It's simple and obvious to spot the telltale characteristics of a healthy catfish. A fish that is fit and healthy will have bright eyes, consume food quickly, and keep swimming around the bottom of the tank. There are a few indications to watch out for when determining whether a catfish is unwell or disturbed. A sick catfish will exhibit warning symptoms include loss of color, lack of appetite, spots, fungus, hazy eyes, bloating, weight loss, irregular swimming, and difficult breathing.

CHAPTER SEVEN

7.1 INSTALLATION OF A CATFISH TANK

While smaller species of catfish may survive in aquariums as little as 5 to 10 gallons (37.85 l), bigger species require at least a 30-gallon aquarium to flourish. Additionally, try to keep the tank free from drafts and harsh sunlight. Catfish, as was previously said, are substrate fish, which means they spend the most of their time at the bottom of the tank and only consume food that reaches that depth. To give the catfish room to look for food, try to cover the bottom of your tank with no more than three inches of gravel. Catfish love to be fed at night when the tank light has been turned off and require water temperatures of between 74- and 78- degrees Fahrenheit (25.56 °C).

Bring your catfish home and set them in the tank while they are still in their bag. This permits them to become acclimated to the water temperature and habitat. Leave them in the bag for 15 minutes before transferring them to the main tank water with a net, leaving the old bag water behind if feasible. Add no more than three catfish to the tank at a time and monitor the chemical balance as you go, as new fish tend to affect the levels.

Tank equipment is best left outside the aquarium; it should be as convenient as feasible. Catfish can transfer the equipment inside if it is there.

Based on the catfish species, the filter must be chosen. Many catfish, including ancistrus, need a strong current, therefore aquariums with these species have considerably faster water flow through the filter than aquariums without these fish. Because many varieties of catfish frequently dig in the ground and raise dirt from the bottom, the filter needs to be strong as well. The water will be turbid since a poor filter won't have enough time to properly perform mechanical cleaning.

1. Tank Decoration

Many plants and shelters should be included in the aquarium. Hollow coconut shells, grottos, caverns, snags, and stumps are all appropriate. Snag will be used as a food source for catfish. They require cellulose; Therefore, they will bite on the snag. Catfish prefer to burrow in the dirt for food. Sand or fine rounded soil is suitable for bottom dwellers. Their skin should not be harmed by the dirt. Don't go overboard with the underwater vegetation - there should be enough room on the ground for all of the aquarium's catfish to lie down on the bottom.

2. Heaters

To keep the temperature comfortable for the fish kept within, most aquariums need some kind of heater. Catfish love water that is between 74- and 78-degrees Fahrenheit (25.56 °C) in temperature. You should spend money on an aquarium heater with enough power to achieve these temperatures. A heater with five watts of power per gallon in your tank is generally what you want to search for. For instance, a ten-gallon tank heater needs 50 watts of electricity, whereas a thirty-gallon tank heater needs 150 watts. In order to maintain a constant temperature across the whole aquarium, larger aquariums can require a heater at either end.

3. Lighting

Catfish require low-lighting conditions. Most catfish are active during the day's darkness and dislike unexpected exposure to light. As a result, it is preferable if the illumination gradually increases and lowers, mimicking the natural cycle of the day. It is advised that an auto-timer be purchased to turn the light on and off.

CHAPTER EIGHT

FAQ AND ADVANCED TIPS

Catfish Farming FAQs

1. What types of catfish are best for farming? Popular choices include channel catfish, blue catfish, and African catfish due to their fast growth rates and adaptability.

2. How do I start a catfish farm? Starting involves selecting a suitable location, preparing a pond or tank, choosing healthy fingerlings, and ensuring proper water quality and feed.

3. What is the ideal water quality for farming catfish? Catfish thrive in clean, oxygen-rich water with a pH between 6.5 and 8.5. Regular testing and maintenance of water parameters are crucial.

4. What do catfish eat in a farming environment? They typically eat commercial catfish feed, which is high in protein, but they can also consume natural foods like insects, algae, and small fish.

5. How long does it take for catfish to grow to market size?
Most catfish reach market

size (1-2 pounds) within 6 to 12 months, depending on the species and farming conditions.

6. What are the common challenges in catfish farming?
Challenges include water quality management, disease control, overcrowding, and finding a reliable market for your fish.

7. Can I farm catfish indoors?
Yes, indoor farming using aquaponics or recirculating aquaculture systems is a viable option, especially in areas with limited outdoor space.

8. **How many catfish can I raise per acre of pond?**
   On average, you can raise 3,000–5,000 catfish per acre in a well-managed pond. Stocking density depends on the availability of oxygen, water quality, and feeding practices.
9. **How do I prevent diseases in my catfish farm?**
   Ensure proper water quality, avoid overstocking, provide balanced nutrition, and quarantine new fish before adding them to your farm. Regular monitoring helps catch early signs of disease.
10. **Can I use natural ponds for catfish farming?**
    Yes, but you need to assess the water quality, ensure there's no contamination, and modify the pond to include proper aeration and drainage systems.
11. **What equipment is needed for catfish farming?**
    Common equipment includes aerators, water testing kits, feeding systems, nets for harvesting, and tanks or ponds for holding fish.
12. **What are the best practices for harvesting catfish?**

Harvesting can be done using seines or traps. It's best to do it early in the morning or late in the evening when temperatures are cooler to reduce stress on the fish.

13. **Is catfish farming profitable?**
Yes, it can be highly profitable with proper management and marketing. Profitability depends on operational costs, scale, and market demand.

14. **Can I combine catfish farming with other fish species?**
Yes, catfish can cohabitate with certain species like tilapia in integrated farming systems, provided their needs are compatible.

15. **How do I market my farmed catfish?**
Market your catfish to local restaurants, grocery stores, and fish markets. You can also sell directly to consumers or explore export opportunities.

Catfish as Pets FAQs

1. **What types of catfish make good pets?**
Popular pet species include

Corydoras, Plecos, and Otocinclus, known for their manageable size and compatibility with aquariums.

2. What size tank do I need for pet catfish?
It depends on the species. Small catfish like Corydoras need a 10-20 gallon tank, while larger species like Plecos require tanks of 50 gallons or more.

3. What do pet catfish eat?
Pet catfish are omnivorous and enjoy a diet of algae wafers, sinking pellets, vegetables, and live or frozen foods like bloodworms.

4. **Are catfish compatible with other fish?**
   Most catfish are peaceful and can live with a variety of fish. However, ensure tank mates are not overly aggressive or too small to avoid predation.

5. **Do pet catfish need a special environment?**
   Yes, they prefer tanks with plenty of hiding spots like rocks, driftwood, and plants. Good water filtration and aeration are essential.

6. **How long do pet catfish live?**
   Depending on the species and care provided, pet catfish can live anywhere from 5 to 15 years.

7. Can catfish clean my aquarium?
   While some catfish species help clean algae and leftover food, they cannot replace regular tank maintenance and cleaning.

8. What common health problems do pet catfish face?
   Issues include fin rot, fungal infections, and stress from poor water quality or overcrowding. Regular monitoring and proper care can prevent these problems.

9. Are catfish easy to breed in home aquariums?
   Some species, like Corydoras, are relatively easy to breed in captivity,

while others require specific conditions, such as changes in water temperature and quality.

10. **Do pet catfish need a heater in their tank?**
Many tropical catfish species require a heater to maintain water temperatures between 72–82°F. Coldwater species, like some types of Plecos, may not need a heater.

11. **Are catfish nocturnal?**
Most species of catfish are nocturnal and prefer to be active during the night. Providing hiding places during the day helps mimic their natural habitat.

12. **Can I keep different species of catfish together in one tank?**
Yes, but ensure the species have similar requirements for water parameters and

space. Avoid overcrowding to reduce stress and competition.

13. **What is the minimum maintenance required for a catfish tank?** Perform weekly water changes (10-20%), clean the tank's filter, and monitor water parameters regularly to ensure a healthy environment.

14. **Can pet catfish jump out of the tank?** Some species, like the pictus catfish, are known to jump. Using a secure lid on your tank helps prevent escapes.

15. **What lighting do catfish prefer?** Catfish generally prefer dim lighting or shaded areas, as they are accustomed to low-light environments in their natural habitats.

16. **Do catfish eat other fish in the tank?** Large catfish species may eat smaller

tank mates, so always consider the size of your catfish and their compatibility with other fish.

17. **How can I tell if my pet catfish is healthy?**

Healthy catfish are active, have clear eyes, smooth fins, and exhibit a good appetite. Signs of illness include lethargy, color changes, or damaged fins.

18. **Can I keep catfish with live plants?**

Yes, but some larger species may uproot plants. Choose sturdy plants like Java fern or Anubias that are less likely to be disturbed.

19. **Do catfish need a specific type of substrate in the tank?**

Many catfish, like Corydoras, prefer soft, sandy substrates that won't damage their sensitive barbels. Avoid sharp gravel.

Catfish Farming FAQs (Advanced)

1. **How do I ensure proper aeration in a catfish pond?** Use mechanical aerators like paddle wheels or diffused aerators to maintain oxygen levels. Position aerators strategically, especially in deeper ponds, and run them during the hottest part of the day when oxygen levels are lowest.

2. **What are the common diseases in farmed catfish, and how do I treat them?** Common diseases include:

- **Columnaris (Cotton Wool Disease):** Treat with antibiotics or potassium permanganate.

- **Ich (White Spot Disease):** Use formalin or salt baths.

- **Parasitic infections:** Manage with copper sulfate or specialized anti-parasite treatments.
  Prevention is key—maintain water quality and quarantine new stock.

3. **What type of feed is best for catfish farming?**
High-protein commercial pellets (28–32% protein) are ideal for growth. Supplement with natural

feeds like insects, earthworms, or agricultural by-products to reduce costs.

4. **Can I use biofloc systems for catfish farming?**
Yes, biofloc technology is highly effective for catfish farming in limited spaces. It enhances water quality by recycling nutrients and reduces feed costs by providing protein-rich flocs.

5. **How do I control predators in my catfish pond?**
Use nets or fencing to protect against birds and animals. Installing scare devices or reflective tapes can deter predators like herons.

6. **What is the best time of year to start farming catfish?**
Spring and early summer are ideal, as warmer temperatures promote faster growth. Avoid starting during winter unless you have temperature control.

Catfish as Pets FAQs (Advanced)

1. **Can I keep catfish in a community tank with other fish?**
Yes, but choose peaceful species as tank mates. Avoid aggressive or territorial fish that may stress catfish. Good companions include tetras, gouramis, and loaches.

2. **What size tank do I need for a large catfish species like the Redtail Catfish?**

Large species like the Redtail require tanks of at least 300 gallons or more. Consider their adult size and ensure sufficient swimming space.

3. **What are the signs of stress in pet catfish?**
Signs include:

- Erratic swimming or hiding excessively
- Loss of appetite
- Faded coloration
- Rapid gill movement (indicating poor oxygen levels)

    Check water parameters immediately if you notice these signs.

4. **How do I acclimate a new catfish to my tank?** Float the sealed bag in the tank for 15–20 minutes to equalize the temperature. Gradually add small amounts of tank water to the bag every 5 minutes for about 30 minutes before releasing the fish.

5. **What are the best water conditions for sensitive species like Corydoras?** Maintain water temperatures between 72–78°F with soft, slightly acidic to neutral water (pH 6.5–7.5). Use a gentle filter to prevent strong currents.

6. **How do I breed catfish in an aquarium?**

- **Corydoras:** Provide soft water, slightly cooler temperatures, and live food to encourage spawning. Females will lay eggs on tank walls or plants.

- **Plecos:** Create caves or tunnels as breeding sites. The male guards the eggs until they hatch.

- **Channel Catfish:** Require a large space with proper hiding areas or spawning boxes.

7. **Why is my pet catfish not eating?**

   Reasons may include:

   - Poor water quality

- Stress from tank mates or new environments
- Illness or parasites
- Overfeeding (leading to uneaten food causing poor appetite)

  Test water, reduce stressors, and offer live or frozen food to entice feeding.

8. **What lighting schedule is best for pet catfish?**
Catfish prefer subdued lighting. A schedule of 8–10 hours of light followed by 12–14 hours of darkness mimics their natural habitat.

9. **Do catfish produce a lot of waste in aquariums?**

Yes, especially larger species. Use a high-capacity filtration system to manage waste and perform regular water changes (20–30% weekly).

10. **Can I hand-feed my pet catfish?**

Yes, many catfish, like Plecos or Corydoras, can be trained to feed from your hand. Use patience and offer food consistently at the same spot.

11. **How long can Catfish survive in a Fish Tank?**

Catfish may live for eight to fifteen years in an aquarium setting if properly cared for. Some species can reach one foot in length at this stage, with the majority of catfish preferring to dwell in groups of three or more. As they prefer to devour smaller species, it is recommended to mix bigger catfish with fish of a comparable size if you decide to get them.

12. How Well Does the Catfish Make a Pet?

Certain animals do make good pets. The Talking, Pleco, Banjo, and Cory Catfish are widely kept as pets. But you should only buy fish that have been reared in captivity. Individuals taken from the wild have the potential to infect your other fish and destroy native populations. The proper species must be chosen while keeping a catfish as a pet. Most of the worries regarding keeping pet catfish come down to size, even if certain species may have been illegally imported and shouldn't ever be kept in your aquarium. All fish increase in size, however certain catfish species are considerably better suited to huge public aquariums than most home aquariums. To guarantee that your catfish will live a long and fulfilling life, be sure to conduct your study on the species you wish to maintain.

13. Is Keeping Catfish as Pets Morally Right?

Quick response: Yes. The length of the response is that it varies on the species and tank size. The primary issue with pet catfish is their capacity for development. All fish begin life as little creatures, although many do not remain that size for very long. No matter what the staff at the pet store may tell you, it is imperative that you conduct your own study and carefully select the kind you want. Discover its mature size to determine the size of the tank it will require.

The amount of space in your aquarium is usually the deciding factor when choosing a catfish species for it.

Farming Techniques

1. Pond-Based Catfish Farming

- **Pond Construction:** Build earthen ponds with a depth of 4–6 feet. Include an inlet and outlet system for water exchange.

- **Stocking Density:** Use 3,000–5,000 fingerlings per acre. For intensive farming, use aeration to increase stocking densities.

- **Feeding Strategy:** Feed juvenile catfish 2–3 times daily with high-protein pellets. Transition to lower-protein feed as they grow. Supplement with organic feeds like rice bran or farm waste.

- **Water Management:** Maintain oxygen levels above 4 ppm and pH between 6.5–8. Regularly remove waste and uneaten food to prevent ammonia buildup.

- **Harvesting:** Use seine nets or draining methods. Harvest when fish reach market size (1–2 pounds) to maximize profitability.

2. **Tank or Cage Farming**

- **Tank Systems:** Use concrete or plastic tanks for controlled farming. Recirculating aquaculture systems (RAS) allow year-round farming with efficient water reuse.

- **Cage Farming:** Place cages in natural water bodies like rivers or lakes. Ensure cages are anchored securely and have good water flow for oxygenation.

- **Advantages:** These systems require less space, provide easy monitoring, and minimize predation risks.

3. Biofloc Technology (BFT)

- **Overview:** A sustainable method where microorganisms recycle fish waste into protein-rich flocs that catfish consume.

- **Setup:** Use tanks or ponds with aerators and a carbon source (e.g., molasses) to support microbial growth.

- **Benefits:** Reduces feed costs, minimizes water usage, and improves fish growth.

Care Tips for Pet Catfish

1. Tank Setup

- **Size:** Choose a tank size based on the adult size of your catfish. Small species like Corydoras thrive in 10–20 gallons, while large Plecos need at least 50 gallons.

- **Substrate:** Use fine sand or smooth gravel to protect their sensitive barbels.

- **Hiding Spots:** Add driftwood, rocks, and plants to create a natural environment and reduce stress.

- **Water Parameters:** Maintain stable conditions with ammonia and nitrite at 0 ppm, nitrate below 20 ppm, and a pH between 6.5–7.5.

2. Feeding Habits

- **Diet:** Provide sinking pellets, algae wafers, and frozen/live foods like bloodworms or brine shrimp.

- **Feeding Schedule:** Feed small amounts once or twice a day. Remove uneaten food to prevent waste buildup.

- **Algae Control:** While some catfish help clean algae,

supplement their diet for balanced nutrition.

3. Common Health Issues

- **Fin Rot:** Caused by poor water quality; treat with antibiotics and improve tank conditions.

- **Barbel Damage:** Results from sharp substrates or poor hygiene. Use sand and clean the tank regularly.

- **Stress:** Avoid sudden changes in water temperature or chemistry and provide sufficient hiding spots.

Breeding Methods for Popular Catfish Species

1. Corydoras Catfish

- **Setup:** Use a breeding tank with soft, slightly acidic water (pH 6.5–7). Add fine-leaved plants or spawning mops for egg-laying.

- **Stimulate Breeding:** Perform a 20% water change with cooler water to mimic rainfall.

- **Process:** Females lay eggs on tank walls or plants, and males fertilize them. Remove the adults to prevent them from eating the eggs.

- **Hatching:** Eggs hatch in 3–5 days. Feed fry infusoria or finely crushed fish food.

2. Plecos (Bristlenose Catfish)

- **Setup:** Provide caves or hollow decorations for spawning. Use slightly alkaline water (pH 7–7.5) with temperatures around 78°F.

- **Breeding Behavior:** The male guards the eggs laid inside the cave. After hatching (4–7 days), fry feed on algae and soft vegetables.

3. Channel Catfish (Farming Breeds)

- **Environment:** Use outdoor ponds or large tanks with water temperatures of 75–85°F.

- **Spawning Habits:** Provide spawning containers like barrels or wooden boxes for egg-laying.

- **Incubation:** After the female lays eggs, the male guards them. Transfer eggs to a hatchery to improve survival rates.

- **Feeding Fry:** Start with high-protein feed or brine shrimp nauplii.

Step-by-Step Guide: Setting Up a Catfish Farm

1. Site Selection and Preparation

- **Choose the Right Location:**

- Select an area with access to clean water, proper drainage, and minimal flooding risk.

- **Prepare the Pond:**
  - Construct an earthen pond with a depth of 4–6 feet.
  - Include inlets and outlets for water flow and install screens to prevent fish from escaping.

2. Water Quality Management

- Test water pH, oxygen levels, and ammonia content regularly using a water testing kit.
- Ideal conditions:
  - **pH:** 6.5–8.5
  - **Oxygen levels:** Above 4 ppm
  - **Temperature:** 75–85°F

3. Stocking Fingerlings

- **Select Healthy Fingerlings:** Look for active, disease-free fingerlings.

- **Acclimatize Fish:**
  - Place fingerlings in the pond slowly to avoid shock from temperature differences.

- **Stocking Density:**
  - Standard density: 3,000–5,000 fingerlings per acre for earthen ponds.

4. Feeding Regimen

- Use high-protein floating feed for juveniles (30–35% protein).
- Transition to grow-out feed (28–30% protein) as fish mature.
- Feed 2–3 times daily, adjusting amounts to avoid overfeeding.

5. Monitoring and Maintenance

- Install aerators to maintain oxygen levels.
- Check for signs of disease or stress (lethargy, discoloration).
- Perform regular water changes and clear waste buildup.

6. Harvesting and Marketing

- Use seine nets for partial harvesting or drain the pond for full harvesting.
- Clean and grade the fish by size before selling.
- Market to local restaurants, fishmongers, or directly to consumers.

Step-by-Step Guide: Breeding Corydoras Catfish

1. Setting Up the Breeding Tank

- Tank size: 10–20 gallons.

- Water conditions: Soft, slightly acidic water (pH 6.5–7) with a temperature of 72–75°F.

- Add fine-leaved plants or spawning mops.

2. Conditioning the Fish

- Separate males and females for 1–2 weeks.

- Feed a high-protein diet (e.g., bloodworms, live or frozen foods).

3. Triggering Spawning

- Perform a 20% water change using slightly cooler water (5–10°F lower).

- Increase oxygen levels with a gentle air stone.

4. Spawning Process

- Females lay eggs on the tank walls, leaves, or mops.

- After spawning, remove adults to prevent them from eating the eggs.

5. Egg Care and Hatching

- Place eggs in a separate tank with gentle aeration to prevent fungal growth.
- Eggs hatch in 3–5 days.

6. Raising the Fry

- Feed infusoria or liquid fry food for the first few days.
- Gradually transition to powdered or finely crushed fish food as they grow.

Advanced Disease Management in Catfish Farming

1. Common Diseases and Treatments

- **Columnaris (Cotton Wool Disease):**

- **Symptoms:** White patches, frayed fins.
- **Treatment:** Potassium permanganate bath (2 mg/L for 1 hour) or antibiotics in feed.

- **Ich (White Spot Disease):**
    - **Symptoms:** White cysts on the skin and fins.
    - **Treatment:** Treat with formalin or a salt bath (1–3 teaspoons per gallon).

- **Aeromonas Infections:**
    - **Symptoms:** Ulcers, hemorrhages.
    - **Treatment:** Use medicated feed containing oxytetracycline or florfenicol.

- **Parasitic Infestations:**

- **Symptoms:** Scratching, gill irritation.
- **Treatment:** Copper sulfate or formalin dip.

2. Preventative Measures

- Regularly monitor water quality and maintain optimal conditions.
- Quarantine new fish for 2–4 weeks before adding them to the main pond or tank.
- Avoid overstocking to reduce stress and disease transmission.
- Vaccinate farmed catfish against common diseases (where available).

3. Emergency Steps for Disease Outbreaks

- **Isolate Affected Fish:** Move diseased fish to a quarantine tank for treatment.

- **Sterilize Equipment:** Use bleach or potassium permanganate to disinfect nets, tanks, and other equipment.

- **Water Treatment:** Perform a partial water change and add appropriate treatments (e.g., salt, antibiotics).

Biofloc Systems for Catfish Farming

1. What is a Biofloc System?

- A biofloc system is a sustainable aquaculture method where beneficial microorganisms recycle waste into protein-rich particles (flocs) that fish consume.

- This system reduces feed costs, improves water quality, and minimizes environmental impact.

2. Setting Up a Biofloc System

- **Tank Design:** Use lined ponds, plastic tanks, or concrete tanks. Ensure a capacity of at least

5,000 liters (1,300 gallons) for commercial setups.

- **Aeration:** Install paddle wheels or air diffusers to provide constant aeration. Proper aeration supports microorganism growth and keeps flocs suspended.

- **Carbon Source:** Add a carbon source like molasses, rice bran, or sugar to maintain a carbon-to-nitrogen ratio (C/N) of 10:1. This promotes the growth of beneficial bacteria.

- **Stocking Density:** Biofloc systems support high stocking densities, ranging from 50–100 catfish per cubic meter.

- **Monitoring:**
    - Maintain dissolved oxygen levels above 4 ppm.
    - Keep ammonia and nitrite levels close to 0.

- Regularly monitor water parameters like pH (6.5–7.5) and alkalinity.

3. Feeding and Maintenance

- Feed fish high-quality pellets in addition to bioflocs. Gradually reduce pellet feed as fish start consuming bioflocs.

- Perform daily checks to prevent floc overgrowth and siphon out excess solids if needed.

4. Advantages of Biofloc for Catfish Farming

- Reduces reliance on external feed by up to 30%.

- Enhances water reuse, making it eco-friendly and cost-efficient.

- Supports year-round farming in controlled conditions.

Advanced Breeding Techniques for Catfish

1. Hormonal Induction for Channel Catfish

- **Why Use Hormonal Induction?**
    - Stimulates spawning in controlled environments where natural conditions are difficult to replicate.

- **Procedure:**
    - Administer hormones like Ovaprim or HCG (Human Chorionic Gonadotropin) via injection. Dosage depends on fish size and species.
    - After injection, place broodstock in spawning tanks with water conditions matching their natural habitat (warm temperatures, soft water).

- **Egg Collection and Incubation:**
  - Collect eggs from spawning boxes or directly from females.
  - Place eggs in hatching trays with gentle aeration.
  - Eggs hatch in 5–7 days at optimal temperatures of 75–85°F.

2. Hybrid Catfish Breeding (Channel x Blue Catfish)

- **Why Breed Hybrids?**
  - Hybrid catfish grow faster, resist diseases better, and exhibit improved feed efficiency.

- **Breeding Process:**
  - Use channel catfish as the female and blue catfish as the male.

- Collect eggs and milt (sperm) manually. Mix them in a container with water to fertilize the eggs.

- **Incubation:**
  - Incubate fertilized eggs in a controlled hatchery with consistent aeration.

3. Fry Care

- **Feeding:** Begin with high-protein starter feed or brine shrimp nauplii.

- **Tank Management:** Maintain clean water with frequent monitoring of ammonia and pH.

Marketing Strategies for Farmed Catfish

1. Direct-to-Consumer Sales

- Sell fresh or live catfish at local markets or through farm outlets.

- Offer custom sizes based on consumer preferences.

2. Supply to Restaurants and Supermarkets

- Partner with local restaurants that specialize in seafood dishes.
- Supply cleaned and packaged fish to grocery stores or supermarkets.

3. Value-Added Products

- Create fillets, smoked catfish, or marinated fish to target premium markets.
- Package fish in vacuum-sealed packs with branding and nutritional information.

4. Online Marketing and Home Delivery

- Set up an online store or partner with food delivery platforms.

- Use social media to promote your farm and products with engaging photos and videos.

5. Export Opportunities

- Explore international markets with high demand for catfish, such as the USA and Europe.
- Meet export standards for quality and packaging.

6. Educational Tours and Farm Branding

- Invite schools, families, or tourists to your farm for educational tours.
- Promote your brand as sustainable, organic, or community-focused to attract eco-conscious buyers.

1. Troubleshooting Biofloc Systems

Common Issues and Solutions

1. **Floc Overgrowth**
   - **Symptoms:** Cloudy water, difficulty in aeration, or fish stress due to low oxygen levels.
   - **Solutions:**
     - Reduce the carbon source for a few days.
     - Use a mechanical filter or siphon out excess flocs to maintain balance.

2. **High Ammonia or Nitrite Levels**
   - **Symptoms:** Lethargic fish, erratic swimming, or red gills.
   - **Solutions:**
     - Increase aeration immediately to

enhance nitrification.

- Perform a partial water change (10–20%) and recheck water parameters.

- Check the C/N ratio; add carbon sources if required.

3. **Low pH Levels**

    - **Symptoms:** Fish stress, poor floc formation.

    - **Solutions:**

        - Add small amounts of lime (calcium carbonate) to gradually stabilize the pH.

        - Avoid sudden pH changes that may harm the fish.

4. **Oxygen Depletion**

- **Symptoms:** Fish gasping near the surface.
- **Solutions:**
  - Immediately increase aeration by adding extra air stones or paddle wheels.
  - Reduce feed temporarily to prevent excess waste buildup.

2. Exporting Farmed Catfish: A Step-by-Step Guide

1. Research Export Markets

- Identify countries with high demand for catfish (e.g., USA, Europe, Middle East).
- Learn about import regulations, food safety standards, and certification requirements.

2. Meet Quality Standards

- **Water Quality Certification:** Obtain certification for farming in clean, pollutant-free environments.

- **Processing Standards:** Ensure fish are processed in HACCP-compliant facilities.

- **Packaging:** Use vacuum-sealed, temperature-controlled packaging with proper labeling.

3. Partner with Exporters or Freight Companies

- Collaborate with logistics companies experienced in shipping seafood.

- Use refrigerated trucks and containers for maintaining freshness.

4. Regulatory Compliance

- Obtain necessary export licenses from your country's trade or fisheries department.

- Prepare documents like:
    - Invoice and packing list.
    - Certificate of Origin.
    - Health Certificate (issued by veterinary or food safety authorities).

5. Marketing in Target Countries

- Build relationships with distributors, seafood markets, and restaurants in the target country.
- Highlight quality, sustainability, and competitive pricing in your marketing materials.

3. Advanced Challenges in Catfish Breeding and Solutions

1. Difficulty in Spawning

- **Challenge:** Broodstock not responding to natural or hormonal triggers.
- **Solution:**

- Check broodstock age (optimal is 2–4 years).
- Adjust water temperature and photoperiod to simulate natural conditions.
- Use quality hormones and ensure proper injection techniques.

2. Low Hatch Rates

- **Challenge:** Eggs not developing or high mortality during incubation.
- **Solution:**
  - Use clean, disease-free water in incubation tanks.
  - Add antifungal agents like methylene blue to prevent fungal growth on eggs.
  - Ensure consistent aeration around egg trays to

prevent oxygen deprivation.

3. Fry Survival Issues

- **Challenge:** High mortality in the first few weeks.

- **Solution:**
    - Feed fry with appropriately sized live food like Artemia or infusoria.
    - Avoid overcrowding; maintain low densities in fry tanks.
    - Monitor ammonia levels closely, as fry are more sensitive to poor water conditions.

4. Advanced Marketing Strategies for Farmed Catfish

1. Digital Marketing Campaigns

- Create a professional website showcasing your farm, processes, and products.

- Use social media platforms like Instagram and Facebook to share engaging photos and videos of your fish and farming process.

- Invest in targeted ads to reach local and international buyers interested in sustainable seafood.

2. Value-Added Products for Premium Pricing

- Develop products like smoked catfish, pre-marinated fillets, or catfish nuggets.

- Package products attractively with a focus on convenience and health benefits.

3. Catfish Subscription Boxes

- Offer monthly delivery of fresh, frozen, or processed catfish to regular customers.

- Include recipes, sauces, or marinades to make the experience unique.

4. Certifications for Credibility

- Obtain organic or sustainable farming certifications to appeal to eco-conscious consumers.
- Highlight these certifications in your branding and marketing efforts.

5. Networking at Trade Shows

- Attend seafood and aquaculture trade fairs to connect with buyers, exporters, and investors.
- Prepare brochures and samples for potential partners.

1. Biofloc Management Guide

Introduction to Biofloc

- **What is Biofloc?**

- A sustainable aquaculture method converting waste into feed.

- **Benefits:**
  - Reduces feed costs.
  - Improves water quality.
  - Promotes eco-friendly practices.

Step-by-Step Guide

1. Setup

- **Tank or Pond Selection:**
  - Use HDPE-lined ponds or large concrete tanks.
  - Capacity: Minimum 5,000 liters for small-scale; larger for commercial.

- **Aeration System:**
  - Install air stones, paddle wheels, or diffusers.

2. Water Quality Parameters

- **Optimal Conditions:**
  - **pH:** 6.5–7.5
  - **Dissolved Oxygen:** > 4 ppm
  - **Ammonia:** < 0.5 ppm
  - **Temperature:** 28–32°C (82–89°F)

3. Biofloc Formation

- Add a carbon source (molasses, rice bran) to balance the C:N ratio (10:1).
- Introduce beneficial bacteria (probiotic powder or natural sources).

4. Feeding and Monitoring

- **Feeding Regimen:** Gradually reduce external feed as biofloces form.
- **Daily Monitoring:**

- Use water test kits for ammonia, nitrite, and pH.
- Observe fish behavior for stress or disease.

5. Troubleshooting Common Problems:

- **Low Floc Formation:**
    - Cause: Low carbon source or insufficient aeration.
    - Solution: Add molasses and check aeration levels.
- **High Mortality Rates:**
    - Cause: Poor water quality or disease outbreak.
    - Solution: Perform partial water changes and use water conditioners.

2. Export Pitch for Farmed Catfish

Objective: Secure partnerships with international distributors and buyers.

1. Key Selling Points

- High-quality farmed catfish raised in sustainable, bio-secure systems.
- Consistent supply of fresh, frozen, or processed products.
- Adherence to international food safety and packaging standards.

2. Target Audience

- Importers in seafood-heavy markets (e.g., USA, EU, Middle East).
- Restaurants, grocery chains, and distributors.

3. Sample Export Pitch

Subject Line: Premium Farmed Catfish – Fresh and Sustainable

Dear [Buyer's Name/Company],

We are thrilled to introduce our premium farmed catfish, raised using sustainable aquaculture practices. Our fish are processed in HACCP-compliant facilities and meet international standards for quality and safety.

Why Choose Us?

- **Unmatched Quality:** Raised in controlled environments, our catfish are rich in flavor and texture.

- **Flexible Supply:** Available in live, fresh, or frozen forms, with customized packaging options.

- **Sustainability:** Eco-friendly farming methods with minimal environmental impact.

We're confident our products can meet your needs for high-quality seafood. Attached, you'll find a detailed brochure and price list.

Looking forward to discussing potential collaboration.

Best regards,
[Your Name]
[Your Contact Information]

3. Marketing Plan for Farmed Catfish

Phase 1: Market Research and Positioning

- Identify demand patterns and preferred product forms (live, frozen, fillets).

- Highlight unique selling points like organic certification or sustainable methods.

Phase 2: Digital Presence

- Build a professional website showcasing your farm, products, and certifications.

- Regularly update social media platforms with engaging content (e.g., cooking videos, farm tours).

- Invest in SEO and targeted ads to attract online buyers.

Phase 3: Value-Added Products

- Introduce items like smoked fillets, marinated catfish, or ready-to-cook packs.

- Offer subscription boxes with recipe cards and bonus items.

Phase 4: B2B Networking

- Attend seafood expos and trade fairs to meet restaurant chains and distributors.

- Partner with local and international delivery platforms for home delivery options.

Phase 5: Feedback and Expansion

- Collect customer feedback to improve products.
- Expand to export-ready markets with strategic collaborations.

Editable Export Pitch Template

Subject Line: Premium Farmed Catfish – Fresh, Sustainable, and Ready for Export

Dear [Buyer's Name/Company],

I hope this message finds you well. My name is [Your Name], and I represent [Your Farm's Name], a premier provider of farmed catfish raised using sustainable and eco-friendly aquaculture practices. We would like to explore the opportunity to supply your company with high-quality catfish that meet the highest industry standards.

Why Choose [Your Farm's Name]?

- **Consistent Quality:** Our catfish are farmed in controlled environments, ensuring rich flavor, texture, and freshness year-round.

- **Sustainable Practices:** We prioritize environmental stewardship by using biofloc systems and reducing waste through efficient water management.

- **Food Safety and Compliance:** Our facility is HACCP-certified, and we adhere to international food safety standards. Our products are also certified by [list any relevant certifications].

We offer the following products:

- **Fresh or Frozen Catfish**

- **Fillets (Skinless, Boneless)**
- **Value-Added Products** (Smoked, Marinated, and Pre-cooked Catfish)

Why Partner with Us?

- **Flexible Supply Chain:** Our production systems allow us to meet demand on a large scale, and we are committed to timely and reliable delivery.

- **Customized Packaging:** We provide packaging tailored to your needs, ensuring freshness and quality during transit.

- **Competitive Pricing:** Our pricing structure is designed to be competitive while maintaining high quality.

Attached is our product catalog, including detailed specifications and pricing. We believe that our farmed catfish would be a great fit for your market, and we would love the opportunity to discuss how we can work together.

Looking forward to your response.

Best regards,
[Your Name]
[Your Position]
[Your Farm's Name]
[Your Contact Information]

Marketing Strategies Tailored to Specific Regions

1. Marketing Plan for the USA Market (Targeting Seafood Retailers and Restaurants)

- **Sustainability Messaging:**
    - Emphasize eco-friendly farming methods, such as biofloc systems and low-impact water management, to appeal to the growing demand for sustainable products.

- Partner with environmental organizations or certifications like "USDA Organic" or "Marine Stewardship Council" to further highlight sustainability.

- **Social Media Campaigns:**

  - Use Instagram and Facebook to showcase your farm, emphasizing the quality of your catfish, its freshness, and the farming process.

  - Collaborate with local chefs or food influencers to create recipes or cooking demos using your catfish.

- **Trade Shows and Seafood Festivals:**

  - Attend seafood industry events like the **Seafood Expo North America** in

Boston to network with wholesalers, retailers, and chefs.

- Offer product samples and provide brochures that detail the sustainability and quality of your products.

- **Retail Partnerships:**

    - Approach major grocery chains like Whole Foods, Safeway, and Costco with sample products, offering competitive pricing and customized packaging.

    - Work with distributors who already have established relationships with these retailers.

2. Marketing Plan for the European Market (Targeting Health-Conscious Consumers and Restaurants)

- **Health and Wellness Focus:**
  - In European markets, where there is a strong emphasis on healthy eating, position your farmed catfish as a healthy, high-protein, low-fat alternative to other fish species.
  - Highlight its omega-3 fatty acids and versatility in European cuisines like Mediterranean and Scandinavian dishes.
- **Certifications and Standards:**
  - Make sure your farmed catfish is certified by relevant European Union regulations, such as **EU Organic** or **ASC (Aquaculture Stewardship Council)**.

- Promote these certifications to ensure transparency and build trust with European consumers who prioritize ethical sourcing.

- **Online Presence and Influencer Marketing:**
    - Launch a website in multiple languages (English, French, German, Italian) with a strong focus on SEO to ensure your farmed catfish is easily found online.
    - Collaborate with European food bloggers and influencers to create content around your farm, your sustainable practices, and your delicious catfish products.

- **Distribution Channels:**

- Focus on establishing relationships with large European supermarket chains like **Carrefour** and **Tesco** and supply fresh, frozen, or value-added catfish.

- Reach out to high-end restaurants offering seafood dishes and pitch your farmed catfish as a premium option for their menus.

3. Marketing Plan for Middle Eastern Markets (Targeting High-End Restaurants and Supermarkets)

- **Exclusivity and Premium Products:**
    - Position your farmed catfish as a premium product for high-end restaurants and affluent consumers in the Middle East. Emphasize its

gourmet quality and versatile use in Middle Eastern dishes like grilled fish, stews, and kebabs.

- **Halal Certification:**
  - Ensure your products are Halal certified to appeal to Middle Eastern consumers. This certification will significantly boost the credibility of your product in the region.

- **Cultural Sensitivity and Packaging:**
  - Use packaging that is culturally appropriate and attractive, potentially incorporating Arabic language labels and imagery that resonates with local tastes.

- **Hospitality and Restaurant Partnerships:**

- Partner with luxury hotels and restaurants in cities like Dubai, Abu Dhabi, and Riyadh to feature your farmed catfish in their menus. Offer them exclusive deals and limited-edition products.

- **Digital Marketing Campaigns:**

  - Leverage digital platforms popular in the Middle East, such as Instagram and Snapchat, for engaging visual content. Feature chefs preparing catfish dishes and share stories about your sustainable farming methods.

THANKS FOR READING

PET CATFISH

www.ingramcontent.com/pod-product-compliance
Lightning Source LLC
Chambersburg PA
CBHW071552220526
45469CB00003B/992